# Cosmology

JOHN GRIBBIN

A PHOENIX PAPERBACK

First published in Great Britain in 1997 by
Phoenix, a division of the Orion Publishing Group Ltd
Orion House
5 Upper Saint Martin's Lane
London, WC2H 9EA

© 1997 John and Mary Gribbin
The moral right of John Gribbin to be identified as the author
of this work has been asserted in accordance
with the Copyright, Designs and Patents Act of 1988

All rights reserved. No part of this publication may be reproduced,
stored in a retrieval system, or transmitted in any form or by any means,
electronic, mechanical, photocopying, recording, or otherwise, without the prior
permission of both the copyright owner and the above publisher of this book.

A CIP catalogue record for this book is available
from the British Library.

ISBN 0 297 81987 9

Typeset by SetSystems Ltd, Saffron Walden
Set in 9/13.5 Stone Serif
Printed in Great Britain by
Clays Ltd St Ives plc

PREDICTIONS

Leeds Metropolitan University
17 0322711 4

## BY THE SAME AUTHOR

In the Beginning
Schrödinger's Kittens
Companion to the Cosmos
In Search of the Big Bang

# Contents

# Chapter 1
# The Birth of the Big Bang

If you look at the sky on a dark, clear moonless night, far away from the dazzle of any city lights, with the unaided human eye you will be able to see no more than a couple of thousand individual bright stars. You will also see a band of hazy light circling the sky, forming the Milky Way; a modest pair of binoculars, or a small telescope like the one built by Galileo in 1609, will show that the Milky Way is made up of a myriad of stars, too faint to be picked out individually by eye. Astronomers estimate that there are at least 100 billion stars in the Milky Way, forming a disc-shaped system so large that light would take 100,000 years to cross it from one side to the other. Those stars are all objects like our Sun – some are bigger, some are smaller, but they are all balls of hot gas, radiating energy that is produced by nuclear fusion reactions (the same sort of process that powers a hydrogen bomb) going on in their hearts. The Sun is an ordinary star, about two-thirds of the way out from the middle of the disc-shaped system of stars that is known as the Milky Way, or as 'our' Galaxy. If each star were the size of a rice grain, the Galaxy would have a diameter the same as the real distance from the Earth to the Moon.

But this impressive island of stars is just one among many. Until well into the twentieth century, it seemed that the Milky Way was the entire Universe. But in 1919 the 100-inch (2.5 metre) diameter Hooker Telescope, the largest built up to then, began operating at Mount Wilson, in California. With this instrument, the American astronomer Edwin Hubble showed in the 1920s that many fuzzy patches of light seen on the sky could be resolved into stars, and that they are, in fact, other galaxies, beyond the Milky Way. Some are disc-shaped, like our own Galaxy; others are spheroidal or ellipsoidal; still others are irregular in shape. As telescope technology has been improved, so that we can see fainter objects further away, deeper into space, more and more galaxies have been found. It is estimated that 100 billion galaxies are in principle visible to our modern instruments, including the Hubble Space Telescope; but only a few thousand of these islands in space have yet been studied systematically.

We live on an ordinary planet, orbiting an ordinary star, in an ordinary part of an ordinary galaxy – the Milky Way Galaxy is, if anything, slightly smaller than the average spiral. There is nothing special about our place in the Universe, and to a cosmologist a whole galaxy, containing hundreds of billions of stars, is merely a test particle, whose motion shows how the Universe as a whole is changing. And it *is* changing. The most important single discovery in the whole of cosmology was also made by Hubble and his colleagues, using the Hooker Telescope in the 1920s. The galaxies are moving apart from one another; the Universe is expanding.

The evidence comes from a phenomenon known as the redshift, an expression that has become part of the language but is often misunderstood. And we know about the redshift because it is possible to analyse the light from distant stars and galaxies, splitting it up into the rainbow pattern of the spectrum in just the same way that light from the Sun can be spread out to make a rainbow pattern using a simple triangular prism.

The rainbow pattern of the spectrum represents a spread of wavelengths of light, from longer wavelengths at the red end of the spectrum to shorter wavelengths at the blue end of the spectrum. It just happens that the way our eyes and brains perceive this part of the electromagnetic spectrum is as a series of colours. There is also much more to the spectrum than we can see; it extends beyond the red end into the still longer infrared and radio wavelengths, and beyond the blue end into the still shorter ultraviolet, X- and gamma-ray regions. To astronomers, all of this is the spectrum, and all of the electromagnetic spectrum is now available for analysis using instruments such as radio telescopes and X-ray telescopes, hoisted (if necessary) above the obscuring layers of the atmosphere on rockets, satellites or balloons. But the essential cosmological redshift was discovered using the visible part of the spectrum and conventional optical telescopes.

What makes spectroscopy so informative for astrophysicists is that each variety of atom (hydrogen, oxygen, iron and all the rest) produces a characteristic pattern of lines in the spectrum. These may be bright lines, if the atoms producing them are hot and radiating energy, or dark lines if the atoms are cold and absorbing energy from light coming from behind them. But they are always produced at precisely the same wavelengths, and each set of lines is as characteristic as a fingerprint, identifying the presence of particular substances in the light from a distant star or galaxy, or in cold clouds of gas and dust in space. It is thanks to spectroscopy that we know what those stars, galaxies and clouds are made of; but in cosmology that is of secondary importance. We know, from spectroscopic studies of different substances on Earth, the precise wavelengths at which these lines are produced. But in the light from distant galaxies the whole pattern of lines is shifted bodily towards the red end of the spectrum – to longer wavelengths. This is the famous redshift; and it tells us how distant galaxies are moving.

When Hubble and his colleagues discovered the cosmological redshift in the late 1920s, astronomers were already familiar with one way to produce this effect. It is called the Doppler effect, and it is caused by motion through space. The Doppler effect is familiar in everyday life from the way the sound of a siren on a vehicle (for example, an ambulance) changes pitch as the vehicle rushes past you. When the vehicle is approaching, the note made by the siren sounds higher, because the sound waves from it are squashed together (making the wavelength shorter) by the motion of the vehicle. As the vehicle passes, the note deepens, because as the vehicle moves away from you the sound waves coming from it are being stretched to longer wavelengths by its motion. In exactly the same way, light from an object moving away from you through space would be stretched, and lines in the spectrum of that light would move to longer wavelengths (towards the red end of the spectrum). It was natural for Hubble to think of the cosmological redshift as a Doppler effect, caused by the galaxies moving outward through space, as if from the site of a great explosion. But this is *not* the cause of the cosmological redshift.

Even before Hubble discovered the cosmological redshift, Albert Einstein had developed his general theory of relativity. This is a theory of gravity, space and time; and since the whole Universe is held together by gravity, the general theory can be applied to give a mathematical description (a so-called model) of the behaviour of the Universe at large. Einstein solved the appropriate equations as early as 1916. This was towards the end of the era when it was still thought that the Milky Way was the entire Universe, and that, although individual stars might be born and die, the system as a whole was unchanging, in the same way that a forest may last for millennia even though each individual tree in the forest lives for only a few decades or centuries. So Einstein was completely baffled when the equations of his new theory insisted that the universe could not be static. The equations told him that according to the general theory of relativity space itself (as part of four-dimensional spacetime, but we don't need to go into that here) must be either expanding or contracting, like a block of rubber being stretched or squeezed. In order to hold everything still, Einstein had to introduce an extra term into the equations, a fiddle factor called the cosmological constant. He later described this as the greatest blunder of his career.

Although Einstein didn't believe what his own equations were trying to tell him, the Russian Aleksandr Friedmann followed up the implications over the next few years, and published, in 1922, his solutions to Einstein's equations, without the fiddle factor and complete with the idea of an expanding Universe. Friedmann had found a family of variations on the theme, all involving expansion. In some models, the universe expands for ever. In others, it expands from a very small size, reaches a maximum size, and then shrinks, possibly 'bouncing' and repeating the whole cycle. In one special case, known as the 'flat' model, it expands ever more slowly, until it hovers for an eternity on the brink of recollapse.

In 1927, unaware of Friedmann's work, Georges Lemaître, in Belgium, made essentially the same discovery, and even suggested that if Einstein's theory were correct the expansion of the real Universe might be discovered by studying the way galaxies moved. Hubble didn't know about any of this work when he discovered the cosmological redshift; but it didn't take long for astronomers to put two and two together and realize the true importance of his discovery.

Although Hubble himself did not know it at first, the redshift had actually been predicted by the general theory of relativity. This is a very powerful reason to believe that the general theory is a good mathematical description of the way the Universe works, and modern cosmology is entirely based on Einstein's theory, which has been tested against ever improving observations of the real Universe and passed every test. What this combination of theory and observations tells us is that the very fabric of the Universe, space itself, is expanding and carrying galaxies along for the ride, like raisins being carried further apart from one another in the rising dough of a loaf of raisin bread. The cosmological redshift is caused by this stretching of space, which stretches light to longer wavelengths on its journey to us from those distant galaxies.

Strictly speaking, we should think of clusters of galaxies being carried apart in this way. Galaxies tend to congregate in clusters, like swarms of bees, held together by gravity. The expanding Universe carries individual swarms apart, producing the cosmological redshift; but within the swarm each galaxy has its own motion through space, like the motion of the stars around our Milky Way. This shows up in the measured redshifts of galaxies in a cluster.

Each cluster has an average redshift, caused by the expansion of the Universe. But each galaxy in the cluster is moving through space, and this motion produces a Doppler effect which shifts its spectral lines slightly (by an amount less than the cosmological redshift). Depending on whether a particular galaxy is moving towards us or away from us, this Doppler effect may decrease or increase the observed redshift, so we see a spread of redshifts around the average for the cluster. There *is* a Doppler shift in the light from distant galaxies, but it is quite independent of, and (except for our very nearest neighbours in space) much smaller than, the cosmological redshift.

The fact that we see the Universe expanding now tells us that it must have been smaller in the past – clusters of galaxies must have been closer together. Using the equations of the general theory of relativity, we can wind the expansion backwards to calculate what the Universe would have been like long ago. We can imagine a time when galaxies touched one another and mingled; a time before that when the stars touched one another and formed a huge fireball; and we can go even further back, using not just our imagination but calculations within the framework of the general theory of relativity, matched against observations of the way the Universe is expanding today. It is this combination of theory and observations which tells us that the Universe must have been born out of a hot, dense fireball (the Big Bang) in the distant past; and the great triumph of cosmology in the forty years following Hubble's discovery of the cosmological redshift was that by the end of the 1960s it could tell us how hot and dense that fireball was, and when the Big Bang occurred.

Apart from revealing the fact that the Universe is expanding, the most important feature of the cosmological redshift discovered by Hubble and his colleagues is that it is proportional to distance. In other words, a galaxy twice as far away from us is being carried away from us twice as fast by the expanding Universe, one three times as far away is receding three times as fast, and so on. This does not mean that we are at the centre of the universal expansion. As it happens, this kind of redshift–distance relation, with redshift proportional to distance, is the only pattern of expansion (apart from no expansion at all) which would look the same from any point in the expanding Universe. It is also the kind of expansion predicted by Einstein's general theory of relativity. *Whichever* galaxy you sit on, you will see the same pattern of behaviour, with the redshifts of other galaxies proportional to their distances from you.

A useful way to picture what is going on is to imagine the surface of a smooth sphere, painted with dots to represent galaxies. If the sphere expands, like a balloon, the distances between the dots will increase, even though the paint blobs do not move through the material the sphere is made of. Suppose that the sphere expands uniformly so that in the time it takes to expand two spots that were 10 centimetres apart become 20 centimetres apart. In that case, two spots that were 20 centimetres apart become 40 centimetres apart, in the same time. The further away a spot is from *any* chosen spot, the faster it seems to recede as the sphere expands. From whichever point you choose to measure, you will see a 'recession velocity' proportional to distance, and there is no centre to the expansion, anywhere on the surface. Except that space has three dimensions and the surface of a sphere has only two dimensions, this is exactly the pattern we see in the real Universe – uniform expansion with no centre.

We know this because it is possible to measure distances to relatively nearby galaxies. Measuring redshifts is easy, and can be done, in principle, for any object we can see, no matter how far away it is. But measuring distances to other galaxies is difficult, and still depends in large measure on a technique used by Hubble in the 1930s. Fortunately, there is a family of stars, called Cepheids, which vary in a regular way. Each of these stars has a brightness related to the period of its variation (the time it takes to brighten, dim and brighten again). There are Cepheids in our own Galaxy, and in its near neighbours, two irregular galaxies called the Magellanic Clouds. So it has proved possible to calibrate the relationship between the real brightness of a Cepheid and its period. This means that when Cepheids are detected in other galaxies (no mean feat even for nearby galaxies, even today, using the Hubble Space Telescope), their periods can be measured, and their true brightnesses calculated. By measuring their apparent brightnesses, this tells us how far away they are, just as you could, in principle, determine the distance to a hundred-watt light bulb by measuring its apparent brightness (or dimness!).

Once you have done this for nearby galaxies, you can determine the redshift–distance relation. And once you have done *that*, you can determine the distance to any galaxy by measuring its redshift and plugging the number in to the known redshift–distance relation.

The snag is that relatively nearby galaxies have such small cosmological redshifts that their random motion through space produces speeds which are a sizeable fraction of their recession velocities (for historical reasons, cosmologists still talk of recession velocities, even though they know the redshifts are not caused by motion through space). This makes it hard to calibrate the redshift–distance relation, because we don't know exactly how much of the redshift is caused by the expansion of the Universe and how much is caused by random motion.

The key parameter is a number now known as the Hubble Constant (H), and in the units used by cosmologists this has a value somewhere between 50 and 70 kilometres per second per Megaparsec. That is, the parameter has a unique value somewhere in that range, but we have not yet been clever enough to pin down exactly where. A Megaparsec is a million parsecs, about 3.25 million light years; if the value of H is indeed 50, that would mean that a galaxy 1 Megaparsec away would have a redshift corresponding to a recession velocity of 50 km/sec, and so on. A widely accepted estimate today is that H is between 60 and 65, and the accuracy of the measurement is likely to be improved dramatically over the next few years, thanks to the Hubble Space Telescope.

Of course, astrophysicists want to pin the number down so they can have accurate measurements of the distances to other galaxies. But cosmologists are more interested in the value of the Hubble Constant for another reason. The number tells us how fast the Universe is expanding. That means it tells you how much time has elapsed since the Big Bang. The bigger the value of H, the smaller the age of the Universe; the smaller the value of H, the bigger the age of the Universe. It is because of the uncertainty in the measurement of H that there is uncertainty in the estimates of the age of the Universe, and all we can say for sure is that between 10 billion years and 20 billion years have elapsed since the Big Bang. My own work, carried out with Simon Goodwin (of the University of Sussex) and Martin Hendry (of the University of Glasgow) in 1997 suggests that the value of the Hubble Constant may be as low as 55 in the usual units, implying that the Universe is at least 13 billion years old. But I shall try to be agnostic about the exact value of the Constant, giving equal weight to other studies.

The widely quoted uncertainty, by a factor of two, in the value of the Constant is sometimes held up to ridicule by people who do not appreciate what a breathtaking achievement it is to measure the age of the Universe even to this accuracy. Certainly, if you only knew the balance of your bank account to within a factor of two you might get into financial difficulties; and an airline pilot who wasn't sure whether the distance from New York to London was nearer 1,500 miles or nearer 3,000 miles would be unlikely to make a success of his profession. But we are talking here about the age of the Universe, the time that has elapsed since the Big Bang. We know that it isn't as small as 1 billion years, and we know that it isn't as large as 50 billion years. We can pin it down to a factor of two, and give as a reasonable round number an age of 15 billion years.

If you stop to think about it, this is an utterly staggering achievement of science, that any generation before the mid-twentieth century would have looked upon in wonder.

What's more, the number we get fits in with other highly significant cosmic ages determined entirely independently. Geologists tell us, on the basis of solid evidence, that the age of the Earth is about 4.5 billion years, comfortably less than the age of the Universe (it would certainly be worrying if the numbers came out the other way around, with geologists telling us that the Earth was 15 billion years old and cosmologists telling us that the Universe was 4.5 billion years old). And astrophysicists calculate that the oldest stars are also around 15 billion years old. To be sure, this does present problems if you prefer the higher values of the Hubble Constant, implying that the Universe is about 10 billion years old. But the way to interpret this 'conflict' is to infer that almost certainly the stellar ages are telling us that the value of H must be low, closer to 50 than to 80, but still in the range suggested by redshift studies.

And, again, it is, in everyday terms, utterly amazing that studies of stars should tell us that the ages of the oldest stars are roughly the same as the age of the Universe. The two numbers are calculated entirely independently, using entirely different techniques. If there was something wrong with our understanding of stars, or our understanding of the Universe, we might have ended up with wildly different 'answers' – stars 100 billion years old in a universe 1 billion years old, perhaps. But we don't. The numbers are in the same ball park, and that gives scientists confidence both in their understanding of astrophysics and in their understanding of cosmology. The remaining differences are significant, and suggest that fine-tuning of their theories is needed; but they suggest that *only* fine-tuning is needed, and that we really do know where we came from, and roughly when.

The Universe was born in a hot fireball some 15 billion years ago. The clinching evidence that persuaded most astronomers that the Big Bang really happened came in the 1960s, when they discovered a faint glow of radiation coming from all directions on the sky. This cosmic microwave background radiation is interpreted as the afterglow of the Big Bang itself – and, astonishingly, it had been predicted twenty years earlier, but the prediction had been forgotten.

The first person to attempt to describe the conditions in the Big Bang quantitatively, using the equations of the general theory of relativity and the best understanding of the behaviour of matter under extreme conditions, was George Gamow (a Russian-born American cosmologist), in the 1940s. Even then, it was obvious that you couldn't take the general theory itself literally as the guide to everything back to the beginning itself, time zero. Winding the present expansion of the Universe backwards mathematically using those equations would imply that it was born out of a mathematical point, a singularity with zero volume and infinite density. This is generally accepted as indicating that there is a flaw in the theory, and that something better (specifically, a quantum theory of gravity) is needed to describe the very earliest stages of the expansion.

Of course, this does not mean that the general theory is wrong in all those applications (and there are many) where it has been tested, and passed the tests with flying colours. Even Newton's theory of gravity is still perfectly good if you want to calculate the load on a bridge, say, and no civil engineer bothers with Einstein's equations. You need Einstein's equations under more extreme conditions, where gravity is stronger, and they also provide a description of the whole Universe. But Einstein's theory includes Newton's theory within itself. A quantum theory of gravity will be able to solve problems involving still greater densities of matter, and puzzles like the birth of the Universe; but it will contain Einstein's theory within itself, as the version that applies under less extreme conditions.

The first people to study the details of the physics of the Big Bang (like Gamow) didn't worry too much about the region close to time zero where the general theory breaks down, but restricted themselves to the period a little later in the history of the Universe, when conditions were less extreme and could be understood in terms of experiments that had been carried out on Earth.

The most extreme form of matter around today is the matter in the nucleus of an atom, the tiny central core, made up of protons and neutrons, that is so dense that a cubic centimetre of nuclear material would have a mass of

100 billion kilograms. The behaviour of matter at such high densities (actually, at rather higher densities) has been studied by experimenters using giant particle accelerators, like those at CERN, in Geneva, and Fermilab, in Chicago.

The behaviour of protons, neutrons and nuclear matter is very well understood, and physicists are confident that they can describe how the Universe at large changed as it expanded away from a state in which the entire Universe had the density of nuclear matter. And that corresponds to a time only one ten-thousandth of a second (0.0001 sec) after time zero itself. As far as the physics is concerned, we understand everything that has happened to the Universe since that time, although working out the details of how a hot fireball of energy with nuclear density produced the galaxies, stars, planets and people we see today is no easy task. It's rather like a game of chess – even a ten-year-old can learn the basic rules of the game, including such subtleties as how the knight moves, and even tricks like castling; but not even a grand master can use those rules to reconstruct the play of a game of chess solely by studying the end game. We know the rules the Universe has operated under since the first ten-thousandth of a second, but we haven't yet used those rules to describe every move in the 'game' that has been played out over the past 15 billion years.

Curiously, it is relatively easy to describe what happened when the Universe was young, and much harder to describe later events (such as the origin of life). This is because a hot fireball is much simpler than a cool Universe. People are complicated systems, in which many different kinds of atom interact with one another chemically in interesting and subtle ways. But if you heat a person up to even a few thousand degrees, all that complexity is broken down and they disappear in a puff of smoke, which is a much simpler substance. Similarly, the Sun is much simpler than the Earth, and the Big Bang was even simpler than the Sun.

So, even back in the 1940s, Gamow and his colleagues were able to calculate how the Big Bang fireball would have cooled as it expanded. They showed that nuclear reactions in the fireball would have produced a mixture of 25 per cent helium and 75 per cent hydrogen in the gas that emerged – exactly the mixture of hydrogen and helium seen (using spectroscopy) in the oldest stars. And they found (updating the numbers slightly) that about 300,000 years after time zero an important change occurred in the Universe.

Before that time, the Universe was so hot that stable atoms could not exist. There were hydrogen nuclei and helium nuclei (each carrying positive electric charge) and there were electrons (each with negative charge), ricocheting around like balls in a crazy pinball machine, in a hot mixture called a plasma. But as the temperature of the whole Universe fell to about 6,000°C, the positively charged nuclei began to cling on to the negatively charged electrons to make electrically neutral atoms. Electromagnetic radiation, including light, can interact with electrically charged particles. So, before atoms formed, the radiation was also involved in the crazy pinball dance, being bounced around from particle to particle. But when atoms formed, there were no charged particles left for it to interact with, and the radiation streamed uninterrupted through the Universe. It decoupled from the matter, and the Universe became transparent.

It is no coincidence that 6,000°C is roughly the temperature at the surface of the Sun; radiation deeper inside the Sun, where it is hotter, is also bounced around between charged particles, to such an extent that it takes 10 million years to get from the core of the Sun to its surface, following a crazy zig-zag path. It only flies freely out into space when it reaches the point where the temperature is low enough for neutral atoms to begin to form. At the time matter and radiation decoupled, when the Universe was 300,000 years old, the entire Universe was in roughly the same state that the surface of the Sun is in today.

Gamow and his colleagues realized that the radiation from the fireball would still fill the Universe today. But because it must have expanded as the Universe expanded, it would have got a lot colder – in effect, being redshifted to much longer wavelengths. It started out, 300,000 years after time zero, with a temperature of a few thousand degrees; using the general theory of relativity, it is straightforward to calculate how much it has cooled as the Universe has expanded. The answer Gamow's team came up with was that the radiation must have a temperature today of a few degrees on the absolute, or Kelvin scale – the scale on which zero is −273°C. Radiation with a temperature of 6,000°C corresponds to orange light, like the light from the Sun. Radiation with a temperature of −270°C corresponds to radio waves in the microwave part of the spectrum, like a very cool microwave oven. Gamow and his colleagues made the prediction that the Universe is filled with this radiation, but did not realize that, even in the 1950s, the technology existed to search for it, using radio telescopes.

In the 1960s, microwave radiation with a temperature of 2.7 K (−270.3°C) was found, by accident, by radio astronomers looking for something else entirely. The radiation comes from everywhere on the sky, and it was soon realized that it was the radiation Gamow had predicted, the afterglow of the Big Bang.

The nature of the cosmic background radiation, which has now been studied in great detail by many instruments (including the COBE satellite) is exactly what we expect from the expanding fireball described by the known laws of physics. Those same laws explain the production of hydrogen and helium in the Big Bang (and also how heavier elements, including the oxygen and nitrogen in your body, were made inside stars later in the life of the Universe), how radiation and matter decoupled after 300,000 years, and why we see that cool glow of background radiation today. There is even a developing understanding of how clouds of gas in the expanding Universe could have been held together by gravity, resisting the expansion, contracting and breaking up to make stars, galaxies and clusters of galaxies. By the end of the 1970s, the Big Bang was firmly established as the best understanding of the Universe.

# Chapter 2
# Cosmology Now

But how did it all begin? Where did that seed of the entire Universe, a hot fireball with the density of nuclear matter, come from? Cosmologists have begun to probe that question in recent decades, and have what they think are some pretty good answers. Not as precise and accurate as their understanding of what happened after the first hundred-thousandth of a second, but intelligent speculations that have made even the subject of the origin of the Universe itself, at time zero, part of serious scientific study, no longer solely the preserve of metaphysicians and theologians. Everything I have told you so far is solid scientific fact, as well established as any theory which attempts to describe what happened long ago and far away can be. What I am about to tell you contains an element of speculation and informed guesswork; but, if anything, that makes it even more intriguing.

There are rival versions of the story of the birth of the Universe, and I shall not go into details of all of them here. But there are two particularly intriguing, and closely related, ideas. All you need to understand the first variation on the theme is a little bit of quantum physics, and a little bit of gravity.

The little bit of quantum physics has to do with quantum uncertainty, a phenomenon discovered in the 1920s by the German quantum pioneer Werner Heisenberg. It turns out that in the quantum world there are certain pairs of related physical properties which can never both be precisely determined at the same time. The classic example is the pairing of position and momentum (which, for our purposes, is synonymous with velocity). We are used to thinking of objects in the everyday world (such as snooker balls) being in a certain place and moving in a certain direction at each instant of time. But in the quantum world, an object such as an electron does not behave in this way. If the position of an electron were precisely known, then it could be moving in any direction at any speed. If the direction and speed with which an electron were moving were precisely known, it would be impossible to say at any instant just where along that path the electron was.

Of course, usually the electron is in a less extreme state, and we can specify both roughly where it is and roughly where it is going; but the more accurately one property is determined, the less precisely the other one is defined.

This is not simply the result of imperfections in our measuring instruments. The uncertainty is built into the laws of physics, so that the electron itself does not 'know' both where it is and where it is going at the same time.

Intriguing though the implications are, this is not the place to go into them. What matters for cosmologists speculating about the birth of the Universe is that there is another pair of these so-called conjugate properties, involving time and energy. It turns out to be impossible to specify the precise energy of a system at a precise time. The more accurately the energy is determined, the more uncertainty there is about exactly when it was determined; the more precise the timing of the determination of the energy, the less certainty there is about the amount of energy.

One astonishing implication of this is that even a region of empty space cannot be said to have zero energy, because that would be a precisely determined amount of energy. Instead, a bubble of energy can flicker into existence out of nothing at all, provided it flickers back out of existence in the time allowed by quantum uncertainty – in effect, before the rest of the Universe has noticed its existence. Because energy and mass are related by Einstein's famous equation $E = mc^2$, this means that for a very brief time even material particles can be created out of nothing at all, provided they disappear very quickly.

It sounds bizarre, but the existence of these so-called virtual particles has a measurable effect on the real particles of the everyday world. In particular, electrically charged particles such as electrons have to be regarded as surrounded by a cloud of virtual particles, and the behaviour of electrons and their interactions with electromagnetic fields can be fully explained only by taking account of the influence of the virtual particles. Each virtual particle exists only for a tiny fraction of a second; but as one disappears another takes its place.

The amount of mass-energy that there is in one of these virtual bubbles of energy determines how long the bubble can survive. The greater the mass-energy, the shorter the lifetime of the bubble – so it is relatively easier to make electrons than it is to make protons, which are much heavier than electrons.

This is where the little bit of gravity comes into the

story. It may surprise you to learn that the total energy of the entire Universe may be precisely zero. This is hard to believe, if you think in terms of all those stars and galaxies, containing vast amounts of mass, and therefore vast amounts of $mc^2$. But all those stars and galaxies are held together by gravity; and gravity turns out to have negative energy. Not just negative energy, but precisely the amount of negative energy required to cancel out the $mc^2$ in all the mass in the Universe.

The full details are best appreciated using the general theory of relativity; but there is a nice physical way to picture how this happens. To start, you have to get hold of the idea that when a cloud of material shrinks under the influence of gravity, gravitational energy is released. This is how a collapsing cloud of gas and dust in space forms a star – it shrinks, and the gravitational energy that is released as a result gets turned into heat, so the collapsing cloud gets hotter. When it is hot enough in the middle (about 15 million degrees centigrade), nuclear reactions begin, and these provide the energy which stops the star collapsing any further, as long as it has nuclear fuel to burn.

The second part of the picture comes from imagining all the material in such a cloud of gas, or in a star, broken down into its component atoms, and dispersed outward in all directions to infinite distance. Because gravity obeys an inverse square law, the strength with which these particles at infinite distance affect one another is zero – the force that matters is proportional to 1 divided by (infinity squared), and that is definitely zero. This means that there is no gravitational energy in the system. But if the cloud of widely dispersed particles is allowed to fall together gradually (obviously, it would need some kind of a nudge to get it going), as it does so energy will be released. You start out with zero energy, and then some energy is taken away. So what you end up with is less than zero.

If you carry the appropriate calculation through fully, using Einstein's equations, you find that if the cloud of particles has a mass $m$, by the time it has contracted to a point it has given out precisely $mc^2$ in energy. In other words, the gravitational energy associated with a mass $m$ is $-mc^2$, equal and opposite to its mass energy.

If you find this hard to swallow, you are in good company. During the Second World War, George Gamow worked as a consultant to the US government. One of his jobs was to take a briefcase full of papers up to Albert Einstein, in Princeton, every couple of weeks. The papers were full of crackpot ideas from inventors who claimed to have devised new wonder weapons that would shorten the war; Einstein, a former patent officer (and a good one), seemed the ideal man to find the flaws in their arguments. The kind of calculation about gravity that we have just described, showing that a star born at a point contains zero energy overall, was first carried out by one of Gamow's colleagues, Pascual Jordan. In his autobiography *My World Line*, Gamow tells how he was walking with Einstein one day, from Einstein's home to the Institute for Advanced Study in Princeton. Gamow casually mentioned to Einstein that Jordon had calculated that a star could be created out of nothing at all, since if all the matter appeared at a point its negative gravitational energy would be numerically equal to its positive rest mass energy.

'Einstein stopped in his tracks,' Gamow tells us, 'and,

since we were crossing a street, several cars had to stop to avoid running us down.'

This idea that stopped Einstein dead in his tracks has now been applied to the whole Universe, not to a single star; some cosmologists see this as the answer to the origin of the Universe itself. It could have been created out of nothing at all, as a quantum fluctuation with zero energy overall. And if a quantum fluctuation lives longer if it has less energy, then a quantum fluctuation with zero energy could, in the words of the song, live forever.

The idea surfaced in the 1970s, but was not taken seriously at the time, because there is one big snag. Obviously, neither the Universe nor a star could be created 'at a point' (a singularity), but that is where you hope quantum gravity will come to your aid, and you assume instead that it might have been created in a state slightly larger than a mathematical point, as a kind of quantum seed, but still much, much smaller than the nucleus of an atom. If you imagine all of the matter in the Universe, all those tens of billions of galaxies, packed together in such a quantum seed, one thing is blindingly obvious. The seed will have an absolutely enormously strong gravitational field, and gravity doesn't blow things apart, it makes them collapse.

At first sight, it seemed that a quantum fluctuation on the scale of the Universe would be as short-lived as any other quantum fluctuation, crushing itself out of existence almost as soon as it was born. But then, in the 1980s, cosmologists found a way to take such a quantum seed and blow it up into the kind of expanding Universe we see around us. The trick is called inflation.

One of the most impressive things about inflation is that the physics behind it comes not from cosmology, but from particle physics. Physicists working at the other extreme of the scale from cosmology, the world of the very small, not the very large, found this mechanism as a natural consequence of their efforts to unite the mathematical description of the forces of nature in one package, a so-called grand unified theory, or GUT.

Like the agreement between the measured ages of the oldest stars and the calculated age of the Universe, this meshing of ideas from the particle world and the world of cosmology is taken as a strong indication that physicists really are on the right track with their theories of how the Universe works. At the end of the twentieth century, particle physics has become part of cosmology, as the theory of the forces and particles of nature has gone beyond the point at which it can be tested by experiments in even the largest accelerators on Earth. The theory deals with energies and temperatures so extreme that in the entire history of the Universe they have existed only in the first split second of the life of the Universe, the interval between the appearance of a quantum seed of superhot energy at superhigh density, and the time, just 0.0001 of a second later, when everything was at the density of an atomic nucleus. The very early Universe is the only laboratory in which the most extreme ideas of particle theory can be tested, and it is tested by seeing (that is, by calculating) how the physics described by those ideas would have shaped the evolution of the Universe, and comparing the predictions of those theories with what we see around us in the real Universe.

Gravity is still the odd one out on this picture, because we do not yet have a complete quantum theory of gravity. And the very beginning of the Universe – the beginning of time – is a mystery. There is a kind of quantum of time, the smallest unit of time that has any meaning, which is called the Planck time, in honour of the quantum pioneer Max Planck. It is ludicrously small – $10^{-43}$ of a second, which is written in decimal notation as a decimal point followed by 42 zeroes and a 1. The best we can say at present is that the quantum seed from which the Universe has grown came into existence – was born, if you like – with an age of $10^{-43}$ seconds, and with gravity already established as an independent force. The size of this primordial seed would have been $10^{-33}$ centimetres, a distance known as the Planck length. So why didn't gravity crush the seed out of existence?

The answer offered by the particle physicists concerns the behaviour of the other forces of nature. There are three of them – electromagnetism, which is a familiar force in the everyday world, and two forces which only operate on a scale smaller than the size of an atomic nucleus, the so-called strong and weak nuclear forces. The weak force is responsible for radioactivity, and the strong force is what holds atomic nuclei together. According to grand unified theory, under the extreme conditions that existed just after the Universe was born all three of these forces were equivalent in strength to one another, and behaved in the same way, as long-range forces. As the Universe cooled, this symmetry was lost, and the forces split apart to take on the characteristics that we know today. The process is called a phase transition, by analogy with the kind of change that occurs when a liquid (such as water) freezes into a solid (in this case, ice).

When water vapour condenses to make liquid water, or when liquid water freezes to make ice, it gives up energy in the form of latent heat, because the state it is changing into stores less energy than the state it is changing from. In an analogous way, phase transitions in the very early Universe released energy, which gave an enormous outward push to the quantum seed, acting as a kind of antigravity and expanding the Universe vastly in a very small interval of time.

The key feature of this expansion is that it was exponential. This means that, while the inflation lasted, in each split second the Universe expanded twice as much as in the previous split second, assuming each split second was the same size. This is like walking down a road in some kind of super seven-league boots. The first step you take covers, say, a metre; the second steps covers two metres; the third step covers four metres; and so on. At that rate, the tenth step would cover more than a kilometre, and the hundredth step would cover more than $10^{30}$ kilometres, which is more than $10^{17}$ light years. Such exponential doubling very quickly runs away with itself.

According to one version of inflation, each doubling in the very early Universe took $10^{-34}$ sec (some versions of inflation suggest even more rapid doubling). But the whole process may have lasted for only $10^{-32}$ sec. It sounds as if it were over in less than the blink of an eye, and it was; but in that brief interval there was time for a hundred doublings. A hundred doublings means that what you end up with is $2^{100}$ times bigger than what you started with. In very round terms, it means that if you start with a quantum seed about $10^{20}$ times smaller than a proton, you end up with a sphere 10 centimetres across, about the size of a grapefruit. At that point, the exponential inflation stops, and the expansion becomes no more than a coast, just as a rocket fired upward stops accelerating when it has used up all its fuel and can only coast upward, gradually being slowed by the tug of gravity; but, once inflation has done its work, the Universe is left expanding so rapidly that gravity will take hundreds of billions of years to halt the expansion. What we see today is a more or less linear expansion, the coasting era, in which each step down the road takes you the same distance as the one before (in fact, each step takes you a tiny bit less far than the one before, because gravity is gradually slowing the expansion). And that is how, many physicists think, a quantum seed smaller than a proton but containing all the mass-energy of all the visible Universe may have got blown up into a grapefruit-sized Universe experiencing a Big Bang. Astonishing though it may seem, this is the new standard model of the early Universe, which forms the basis of most current research in cosmology. But that still leaves intriguing questions for the next generation of cosmologists to tackle.

Chapter 3

# Into the Future

If you don't like the idea of the primordial quantum seed appearing out of nothing at all, there is another way in which it could have been born, one which some cosmologists find even more appealing. This is the idea that our Universe is just one among many, and that there was no unique 'beginning', just an interconnected web of universes extending forever in both time and space. The way to get a handle on this idea is to think about what may happen inside a black hole in our Universe. Now, we are entering deeper into the realms of informed speculation, and some of the ideas I will discuss are by no means fully worked out scientific theories. They are not even, strictly speaking, predictions. Instead, they are signposts into the future, pointing the way for researchers who may develop those fully worked out theories in the decades ahead.

A black hole is a place where matter collapses indefinitely, under the influence of gravity, towards a singularity. It is a mirror-image in time of the way in which the Universe expands outwards from something close to a singularity, and is described by exactly the same equations, but with the direction of time reversed. We can only say that the matter falling into a black hole heads 'towards' a singularity, not into a singularity, because just as with the birth of the Universe we expect quantum gravity to become important very close to the singularity, at about the Planck length.

Without a complete quantum theory of gravity, it is impossible to say exactly what happens in the heart of a black hole. But two researchers based in the United States, Lee Smolin and Andrei Linde, have independently come up with the idea that some kind of 'bounce' may occur, turning the collapse into an expansion, and shunting the material falling in towards the singularity sideways into a new set of dimensions – its own space, and its own time. It is as if the black hole is the entrance to a tunnel through what the science fiction writers call hyperspace. Less elegantly, the cosmologists call such a tunnel a wormhole, and suggest that it connects to another region of spacetime (another universe), in which the matter from the black hole blasts outward in just the same way that matter burst out from close to a singularity at the birth of our own Universe.

And they do mean *just* the same way. Remember that if you start from a singularity, negative gravitational energy cancels out the positive mass energy of whatever it is that comes out. So you could make a black hole with, say, ten times as much mass as there is in our Sun, and what would emerge at the other end of the wormhole would not be a tiny universe containing only ten solar masses of matter, but a full-sized universe, containing perhaps as much matter as our own – or more. The few solar masses that go into the black hole in the first place are simply a catalyst, needed to trigger the formation of the singularity.

If this can happen to matter that falls into a black hole

in our Universe (and all the evidence is that it can), then our Universe could have been born in the same way, from matter falling into a black hole in some other region of spacetime. It is possible that the overall structure of space-time (sometimes called the 'Metaverse') is a series of interconnected bubbles, resembling froth on a glass of beer, with no beginning and no end.

Another way to picture this is to go back to the analogy between the expanding Universe and the skin of an expanding balloon. A baby universe would correspond to a piece pinched off from the skin of 'our' balloon, making a little blister which then begins expanding in its own right, with new baby universes budding off from its own skin, and so on indefinitely.

Strange though these ideas are, and they are admittedly speculative, they represent the serious speculations of respectable cosmologists today, and they are about as well developed as the idea of the Big Bang itself was in the 1940s. As we stand today, it seems clear that some form of inflation happened to put the Universe into the state that is generally referred to as the Big Bang; but nobody can say for sure just what it was that got inflation going. What we do know is that inflation is hugely successful in explaining the overall appearance of the Universe today.

There are many curious features about inflation. For a start, it seems to take place faster than the speed of light. Even light takes a little over 3 billionths of a second ($3 \times 10^{-9}$ sec) to cross a single metre, but inflation expands a region many times smaller than a proton to become a sphere 10 centimetres across in only about $10^{-32}$ sec. This is possible because nothing is moving through space –

space itself is stretching, like a stretching rubber sheet, and taking matter along for the ride.

This rapid expansion predicts many features that are seen in the real Universe, notably its extreme uniformity. The Universe is, in spite of the presence of galaxies and clusters of galaxies, a remarkably uniform place, and this is most clearly seen by looking at the background radiation, which tells us what the Universe was like 300,000 years after the Big Bang. Remember that the average temperature of the background radiation is 2.7 Kelvin. COBE found that there are tiny variations in the temperature of this radiation coming from different parts of the sky, amounting to fluctuations of about thirty-millionths of a degree. That is an indication of just how smooth and uniform the Universe was at that time; the irregularities were no bigger than one part in a hundred thousand, or 0.001 per cent (and, incidentally, those fluctuations are just big enough to have grown into clusters of galaxies in the time available).

This uniformity in the early Universe is explained because everything we can see today has come from such a tiny seed, a quantum fluctuation so small that there was no room inside it for any irregularities. Even better, the ripples in the background radiation revealed by COBE have exactly the right structure to be explained as coming from further quantum fluctuations that occurred during the era of inflation, that got frozen in and stretched hugely by the inflation process, but originally occurred when what is now the entire visible Universe was something like $10^{-25}$ centimetres across – 100 million times bigger than the Planck length. All the irregularity in the Universe today (and that includes us) has, on this picture, grown from these tiny quantum ripples.

Inflation can also explain another remarkable feature of the Universe, the fact that it sits very close to the dividing line between eternal expansion and eventual recollapse. The analogy with the rocket being fired upward from the surface of the Earth comes in useful once again here. Inflation accelerated the expansion of the Universe, but when inflation stopped the expansion began to coast, slowing down under the tug of gravity. In much the same way, a rocket accelerates upward until all its fuel is gone, and then coasts, slowing as it is tugged back by gravity. If the rocket is going fast enough when it runs out of fuel, it will escape from the Earth and go out into space. But if it has less than this critical escape velocity when it stops accelerating, gravity will eventually halt its rise and then bring the rocket crashing down to Earth.

The Universe faces a similar pair of possibilities. If it is expanding fast enough, then gravity can never halt the expansion, which will continue for ever. If it is expanding at less than the critical speed, gravity will one day halt the expansion, then reverse it, eventually crushing everything back into a state like the Big Bang from which the Universe emerged – a Big Crunch, which might conceivably 'bounce' into a new Big Bang. And if the Universe is expanding at exactly the critical speed, gravity will just be able to halt the expansion, but will take for ever to do so.

Which fate the Universe faces depends on only two things – how fast it is expanding (which can be measured from the redshift) and how much matter there is in the Universe trying to pull itself together by gravity. If the

Universe will expand for ever, it is said to be 'open'; if it will one day recollapse, it is said to be 'closed'. And if it sits exactly on the dividing line between being open and being closed it is said to be 'flat'. The names come from the way space itself is curved by gravity in the different scenarios. A closed universe is the three-dimensional equivalent of the closed surface of a sphere, like the surface of the Earth. An open universe extends for ever, curving outward like the open surface of a mountain pass, or a saddle. And a flat universe is the three-dimensional equivalent of the flat surface of a piece of paper.

Since we know how fast the Universe is expanding, the only thing we need to measure to determine its fate is how much mass it contains. Fortunately, because the Universe is so uniform, we don't actually have to add up all the mass, but just find the density in a representative large volume of space (large enough, that is, to include several clusters of galaxies). The critical density needed to make the Universe flat can be calculated with great precision – all that is needed is somewhere between $10^{-29}$ and $2 \times 10^{-32}$ grams per cubic centimetre, averaged over the entire Universe. Since a single hydrogen atom weighs $1.7 \times 10^{-24}$ grams, this means that on average, for the Universe to be flat, it must contain a single hydrogen atom in every hundred thousand cubic centimetres of space, or ten hydrogen atoms in every cubic metre. If all the bright stuff in all the stars and galaxies we can see were smeared out evenly through the Universe, it would contribute only about a hundredth of this critical density – but that is not the whole story.

The density of the Universe is usually described in terms

of a parameter called omega, in such a way that if omega is 1 the Universe is flat, if omega is less than 1 the Universe is open, and if omega is greater than 1 the Universe is closed. It is difficult to measure omega precisely, mainly because we cannot see all of the matter in the Universe. Obviously, as well as the bright stars and galaxies there must be some dark stuff, but how much? To some extent, we can estimate how much dark stuff there is by studying the way galaxies move. In a cluster of galaxies, the spread of redshifts tells us how quickly the galaxies are moving relative to one another. The only thing that can hold such a cluster together and stop it flying apart is gravity, and the faster the galaxies are moving the stronger the gravitational pull needed to hold the swarm together – which means more dark matter in the cluster to provide the gravity. These studies show that there is at least ten times as much dark matter in large clusters of galaxies as there is bright stuff; what we can see is only the tip of the proverbial iceberg.

So there is no doubt that the Universe contains at least 10 per cent of the critical density of matter needed to make it flat, but even allowing for the possibility of more dark matter that has not yet been detected it seems extremely unlikely that it contains more than the critical density. At first sight, this looks like one of those ropy cosmological estimates – an uncertainty of a factor of 10 in the measurement of a key cosmological number! Even worse than the uncertainty of a factor of 2 in estimates of the age of the Universe. But, like those age estimates, on closer inspection it turns out that this is a very significant discovery.

Unless you can think of some good reason otherwise, it

would seem that the Universe could have emerged from the Big Bang with any expansion velocity, and there would then be no reason for the density of the Universe today to be anywhere near the crticial value – which is the only cosmological density that has any special significance. It could have been one-thousandth, or one-millionth of the critical value – or a million times as big.

You can see just how truly remarkable the actual measured density of the Universe is by looking at how the relationship between density and expansion rate changes as the Universe expands. Imagine an open universe emerging from the Big Bang with nearly the critical density appropriate at that time. An open universe expands relatively rapidly, at more than the critical velocity, so it thins quickly, and the density drops rapidly, reducing the effectiveness of gravity in slowing the expansion and increasing the discrepancy between the density needed to close the universe and its expansion rate. As the universe expands and the density decreases, at succeeding epochs the density will be further and further away from the critical value appropriate at those epochs. The same thing happens in the opposite direction if the universe starts out from the Big Bang with even a little more than the critical density. Gravity slows the expansion of the universe rapidly, so that even though the density decreases as the universe expands, the expansion rate decreases more, and the grip of gravity is relatively stronger in succeeding epochs, so the universe becomes more and more obviously closed.

So if the density of the real Universe today is within a factor of 10 of the critical value, even though the Universe has been expanding for some 15 billion years, further back in time (closer to the Big Bang) it must have been closer to the critical value. How close? If you carry the calculation through accurately, you find that at the end of the era of inflation, at the beginning of the Big Bang the Universe must have been flat to a precision of 1 part in $10^{60}$. Far from the present measurements of the density parameter, to an accuracy of a factor of 10, being an embarrassment to cosmologists, this means that the value of the density parameter in the Big Bang is actually the most accurate observationally determined number in the whole of science!

Because the critical density itself is the only density that has any special significance, while 10 per cent of the critical density certainly does not have any special significance, this remarkable fact encourages many cosmologists to speculate that the Universe does indeed have precisely the critical density, and is indeed flat.

This idea is saved from being mere speculation by inflation. One of the side effects of inflation is that whatever the curvature of space you start out with, you end up with a very nearly flat universe. Imagine starting out with something like the shrivelled, wrinkly surface of a prune. When a prune is placed in water, it swells up, and all the wrinkles are smoothed out to make a round surface – but still one that is definitely closed, in the sense used by cosmologists. But now imagine doubling the size of that prune a hundred times or more, the equivalent of the inflation that happened in the very early Universe. A hundred doublings would take a sphere a couple of centimetres across and blow it up to a diameter of more than a thousand billion light years, roughly a hundred times the size of the visible Universe. You would still have a surface that was technically closed, forming a sphere; but it would be so huge that any creatures living on the surface of the sphere would be unaware of the curvature, and every test they could devise would show that the surface was flat. Inflation theory suggests that our Universe is like that. It may actually be just closed (or, perhaps, just open), but the actual value of the density parameter omega is so close to 1 that no human observations will ever be able to measure the difference.

Because there is still no direct evidence for the presence of all the dark matter that would be needed to make the Universe flat, some cunning cosmologists have devised variations on the inflationary theme that would allow a universe to emerge from inflation with slightly less than the critical density. Although the ingenuity of these ideas is admirable, the wise money today is betting on the likelihood that the Universe really is flat. And this has led to a wave of experimental activity and new links between the cosmologists and the particle physicists, as they join forces to search for the dark matter that makes up the bulk of the Universe.

The bright stuff that we can see in stars and galaxies makes up only about 1 per cent of the critical density. The way galaxies move in clusters shows that there is at least ten times as much dark stuff, but even this still only adds up to about 10 per cent of the critical density. For the Universe to be flat, there must be ten times more dark matter, and there is independent evidence for this from the way whole clusters of galaxies are seen to be streaming across space, tugged on by some invisible mass. All this means that all the bright stuff we see in stars and galaxies is outweighed by dark matter 100 to 1. The reason why the particle physicists are excited by this possibility is that our understanding of the Big Bang tells us that most of this dark matter cannot be in the same form as the stuff that stars, galaxies, planets and people are made of.

One of the great triumphs of the standard Big Bang model, dating from the 1960s and resting on the secure foundation of everyday physics involving no conditions more extreme than those of nuclear matter today, is that it predicts the mixture of materials emerging from the Big Bang to form the first stars – roughly 75 per cent hydrogen, 25 per cent helium, and the merest smattering of other light elements such as deuterium. These calculations were first carried out approximately by George Gamow and his colleagues in the 1940s, and with great accuracy by Fred Hoyle and his colleagues in the 1960s. The agreement between what the theory predicts and what observers see in old stars, which are presumably made of primordial material, is so striking that there can be no doubt that the Big Bang is a good description of what went on in the Universe from about 0.0001 sec after time zero to about 4 minutes after time zero (of course, the agreement between observation and theory is not perfect, but the differences suggest that no more than fine-tuning is needed to make it better).

But there is another implication of all this. The calculations depend on the rate at which nuclear reactions were going on in the Big Bang, chiefly converting hydrogen into helium. This depended on the rate at which the Universe was expanding and cooling (which we know from winding back the observed expansion, and from the temperature of the background radiation today), the temperature, and the density of the material taking part in those nuclear reactions. Nuclei are made of protons and neutrons, and protons and neutrons are members of a family of particles called baryons, so nuclear matter (and, by extension, everyday matter including the stuff you and I are made of, the Earth, Sun and everything we can see in the Universe) is called baryonic matter. The calculations of how nuclear reactions went on in the Big Bang, which agree so well with observations of the composition of the oldest stars, say that the density of baryonic matter emerging from the Big Bang was no more than enough to provide a few per cent of the critical density needed to make the Universe closed today. Even by pushing all the numbers to the limit, the amount of baryonic matter in the Universe today cannot be more than 10 per cent of the critical density. Five per cent is regarded by most astronomers as a much more realistic upper limit.

Because we are made of baryons, for many years astronomers were influenced by a kind of unconscious baryon chauvinism into assuming that this meant that the *total* density of the Universe could not be more than a few per cent of the critical density. It was only when studies of the way clusters of galaxies move showed the influence of large amounts of dark matter in the Universe, and inflation suggested that the Universe must be extremely flat, that, in the 1980s, the penny dropped, and cosmologists began to investigate the possibility that there might be other kinds of matter in the Universe, non-baryonic dark stuff which did not participate in those crucial nuclear reactions during the Big Bang.

As soon as they took on board the idea of non-baryonic dark matter, astronomers realized that it could help to solve one of the great puzzles of the Universe – how galaxies formed. Although the dark stuff (or at least most of it) had to be non-baryonic, it would still interact with other matter by gravity. In the expanding Universe, if the density today really were only 10 per cent of the critical density, it would be hard (indeed, impossible) to see how structures as large as galaxies and clusters of galaxies could have formed from primordial fluctuations, by tugging clouds of gas together gravitationally, even in the 15 billion years that have elasped since the Big Bang. But adding in ten times as much non-baryonic dark matter provides just the right amount of extra gravitational pull to do the job.

The way galaxies would have formed under these circumstances can be simulated in a computer, and the simulations look very like the real Universe. Again, as with the match between the composition of the oldest stars and the calculations of the mix of baryons that emerged from the Big Bang, the agreement between calculation and observation is not perfect. But, again, the discrepancies are so small as to suggest that only fine-tuning is needed to make the agreement better.

In order to make their discussions as general as possible, the cosmologists usually make no assumptions about just what the dark matter might be, but refer to it in descriptive terms. There are two forms that the dark matter might take, assuming that it is made up of particles produced in

LEEDS METROPOLITAN
UNIVERSITY
LEARNING CENTRE

the Big Bang and spread more or less evenly through the Universe. Hot dark matter (HDM) would consist of very light particles, much lighter even than an electron, which stream through the Universe at high speeds, close to the speed of light. Cold dark matter (CDM) would consist of more massive particles, each perhaps even more massive than a proton, which travel through the Universe at low speeds. Together, both kinds of dark matter are sometimes referred to as WIMPs, a rather tortuous acronym for weakly interacting massive particles, a name intended to indicate the key properties of the particles, that they have mass (and therefore interact with other matter gravitationally) but scarcely interact in any other way at all.

All of this was great news for the particle physicists, because those same grand unified theories that indicate the possibility of inflation also predict that there must be as yet unidentified varieties of non-baryonic particles in the Universe – both CDM and HDM. Once again, both cosmology, the study of the Universe on the largest scale, and particle physics, the study of the world of the very small, are suggesting the same thing about the Universe – in this case, that dark matter, in the form of WIMPs, should exist.

The physicists already knew about one kind of particle that could contribute to the non-baryonic dark matter. They are called neutrinos, and they are involved in nuclear reactions, so their presence is allowed for in calculations of the reactions that took place in the Big Bang. But it was originally thought that neutrinos had zero mass, so although the Universe was known to have been full of them, it had been thought that they had no gravitational influence, and could not contribute to slowing the expansion of the Universe.

Neutrinos are extremely reluctant to interact with baryonic matter, even though they are produced in profusion in nuclear reactions. Their existence was first predicted by theorists in the early 1930s, but they are so difficult to detect that it wasn't until the mid-1950s, a quarter of a century later, that they were first identified, pouring out from a nuclear reactor at the Savannah River site in the United States. We now know that the Universe is filled with neutrinos, and a great flood of them washes through us from the Sun, which is, of course, supported by the energy released by nuclear reactions going on in its heart. But they will do you no harm; neutrinos pass through a human body, or even the solid Earth, more easily than light passes through a pane of clear glass. If a beam of neutrinos like those produced inside the Sun were to travel through solid lead for a distance of a thousand parsecs (more than 3,000 light years), only half of them would be captured by the nuclei of lead atoms along the way.

This reluctance of neutrinos to interact explains why even in the 1980s nobody could be quite sure whether or not they have a tiny mass. The electron is the lightest component of ordinary atomic matter, and weighs in with half of one-thousandth of the mass of a proton – so light that its gravitational influence on the Universe is usually lumped in with the protons and neutrons. If neutrinos do have mass, experiments show that it could be no more than about one-hundredth of 1 per cent of the mass of the electron, which for most purposes can indeed be regarded as zero. But there are a *lot* of neutrinos in the Universe, several billion for every baryon, left over from the Big Bang. So many, in fact, that, as astronomers realized in the 1980s, even if each one has only a tiny mass, their overall gravitational influence could greatly exceed the influence of all the baryons put together (but don't imagine that the neutrinos pouring out from stars like the Sun are adding to the density of the Universe; those neutrinos are made out of the mass energy of the stars themselves, helping to compensate for the loss in mass of the stars as $mc^2$ is converted into $E$).

Neutrinos are the archetypal form of hot dark matter, zipping through the Universe (including right through stars, planets and people) at close to the speed of light. But computer simulations show that, if all the dark matter were in the form of neutrinos (or any kind of HDM), galaxies and clusters of galaxies like the ones we see in the real Universe could not form, because the fast-moving particles would tend to blow structures apart as they started to form, like a cannonball demolishing a wall made of loosely piled bricks. This is not a problem today, because baryonic matter is spread fairly thinly across the Universe, and is almost transparent to neutrinos; but it would have been a problem during the first few seconds of the Big Bang, when the density of the Universe was very high (greater than the density inside a star today), and the neutrinos would have smoothed out any irregularities imprinted by inflation.

The computer simulations show that the best representation of the pattern of galaxies and clusters of galaxies seen in the real Universe comes from calculations in which there is a mixture of roughly one-third hot dark matter, two-thirds cold dark matter, and a smattering (a few per cent) of baryonic matter. Slow-moving, massive CDM particles tended to clump together gravitationally very early in the life of the Universe, producing gravitational 'pot-holes' which attract baryonic matter. The baryonic matter (mainly hydrogen and helium) falls into the pot-holes under the influence of gravity, and eventually forms galaxies. But if all the dark matter were in the form of CDM, this process would be too efficient – the resulting pattern of galaxies and clusters of galaxies would be more concentrated than we see in the real Universe (of course, the computer simulations do not reproduce the actual pattern of clusters seen today, but a general pattern in which it is possible to measure features such as the numbers of clusters of different sizes, and their separation from one another, and compare these with the equivalent features of the pattern of galaxies in the real Universe). So the favoured model today is mixed dark matter, or MDM.

The dramatic implication of all this is that roughly two-thirds of all the matter in the Universe is not just dark, but is in the form of varieties of particles (perhaps just one kind of particle, perhaps several) that have never been detected on Earth, but which are predicted to exist (and were predicted even before astronomers realized the need for dark matter in the Universe) by particle physics theory. The particle physicists have no qualms about describing the properties of these hypothetical particles in detail, and giving them names, such as axions and gravitinos.

The main problem is that different versions of the grand unified theories predict different varieties of dark matter particles, so there are more candidates than we need to do the job of flattening the Universe. But the positive side of this is that, if some of these particles can be identified and their properties measured, that will tell us which versions of the GUTs are correct, and will eliminate the other possibilities.

Catching a CDM WIMP is not easy, because, like neutrinos, they are reluctant to interact with everyday matter. But, because they have much more mass than a neutrino, and fill the Universe, it may not be all that difficult, either. One kind of possible CDM particle would have a mass comparable to that of a proton, and on average there would be one such particle in every 5 cubic centimetres of space, if the Universe is flat. I don't just mean every 5 cubic centimetres of 'empty' space, but every 5 cubic centimetres in the Universe, including inside the Sun, and passing through the room you are sitting in and through your body as you read these words. Each litre of air that you breathe contains a few hundred WIMPs, on this picture.

The way to detect such a WIMP is to monitor the occasional collision between the CDM particle and an ordinary atom – strictly speaking, the nucleus of an atom. A WIMP detector (and there are several now operating around the world) consists of a lump of material, a small crystal cooled to a temperature of a few Kelvin (about the temperature of the background radiation), which is monitored for any sudden rise in temperature caused by the impact of a WIMP, converting the energy of motion of the WIMP into heat. The impact of the kind of CDM WIMP I have described will raise the temperature of the crystal by a few thousandths of a degree.

Nobody has yet reported the detection of such an event,

but it could happen any time in the next few years. Physicists interested in finding the two-thirds of the Universe that has not yet been seen are even now busily searching for it, not with the aid of giant telescopes peering outward into space, but with the aid of small, supercooled crystals being monitored in underground laboratories, far from any potential source of interference.

And there is another way to search for dark matter. The distribution of matter in the very early Universe, the fireball of the Big Bang itself, must have left an imprint on the cosmic microwave background which still fills the Universe today. The detailed structure of these ripples in the background radiation depends on how much matter there was in the Universe at that time (and therefore, of course, on how much matter there is in the Universe still). This structure is on a finer scale than the instruments on the COBE satellite could detect, but its structure can be predicted by different versions of cosmological theory, with different amounts, and different kinds, of dark matter. The next generation of satellites designed to probe the background radiation will be launched in the early years of the twenty-first century, and should soon be able to measure the density of the Universe to an accuracy of 1 per cent. By the year 2010, we will know, for sure, exactly how much dark matter there is, and what proportion of it could only be accounted for by the presence of WIMPs. There is a very real race on today between the astronomers searching for traces of WIMPs, and the particle physicists searching for WIMPs themselves.

Those next-generation satellites will also be able to measure the Hubble Constant to an equivalent accuracy, providing the answers to questions that cosmologists have worried over for decades. Within twenty years, it might seem as if cosmology, as envisaged in the 1960s and 1970s, had been completed, with every *i* dotted and every *t* crossed. But don't think that this will be the end of this branch of science. Already, a few informed speculators are laying down markers for the way things may develop further into the twenty-first century. The most intriguing of these ideas comes from Lee Smolin, who has developed the concept of baby universes, being 'born' (or budded off) from their parents through black holes, to suggest that a form of Darwinian evolution may be at work as this process unfolds. He suggests that there may be slight variations in the laws of physics from one generation of universes to the next, mutations which allow universes to compete with one another for the opportunity to expand and grow, and to make more baby universes, in the immensity of multi-dimensional spacetime.

This idea, a marriage between the Einsteinian theory of relativity and the Darwinian theory of natural selection, is far from being widely accepted, and causes irritation to many cosmologists. It may turn out to be wrong. But it is likely to provide a paradigm which will encourage astronomers to think up new tests and ways to probe the Universe, and even by proving it wrong, those tests would undoubtedly give new insights into how the Universe works. If they proved it right, it would be the biggest revolution in science since Copernicus displaced the Earth from the centre of the Universe.

This is the cutting edge of cosmology today, where informed speculation combines with observation and experiment to improve our understanding of the Universe. But I don't want to leave you with speculation, no matter how respectable. Instead, I want to leave you with solidly established fact, the standard model of the Big Bang, which describes the evolution of the Universe from a time one-hundred-thousandth of a second after time zero. There is nothing speculative in the summary that follows, describing the first 300,000 years in the life of the Universe – that is, the epoch immediately *after* the first ten-thousandth of a second. The science behind this understanding of the early stages in the life of the Universe is arguably the greatest achievement of the human intellect, and it is almost as remarkable that it can be summed up in a few hundred words. As Albert Einstein once said, 'the eternal mystery of the world is its comprehensibility . . . the fact that it is comprehensible is a miracle'.

# Epilogue
# The Great Achievement

The standard model of the Big Bang tells the story of everything that has happened since 0.0001 ($10^{-4}$) of a second after the moment of creation. At that time, the temperature of the Universe was $10^{12}$K (1,000 billion degrees) and the density was the density of nuclear matter, $10^{14}$ grams per cubic centimetre (1 gram per cubic centimetre is the density of water).

Under these conditions, the photons (particles of light) of the 'background' radiation carry so much energy that they are interchangeable with particles, in line with Einstein's equation $E = mc^2$. Photons create pairs of particles and antiparticles, such as electron–positron pairs, proton–antiproton pairs, and neutron–antineutron pairs, and these pairs annihilate one another to make energetic photons in a constant interchange of energy. There were also many neutrinos present in the fireball. Because of a tiny asymmetry in the way the fundamental interactions work, slightly more particles were produced than antiparticles – about a billion and one particles for every billion antiparticles.

When the Universe cooled to the point where photons no longer had the energy to make protons and neutrons, all the paired particles and antiparticles annihilated, and the one in a billion particles left over settled down as stable matter.

One-hundredth of a second after time zero, with the

temperature down to 100 billion K ($10^{11}$ K), only the lighter electron–positron pairs still interacted in the dance with radiation. Protons and neutrons had settled out of the maelstrom. At that time, there were as many neutrons as protons, but as time passed interactions with energetic electrons and positrons shifted the balance steadily in favour of protons. One-tenth of a second after time zero, the temperature was down to 30 billion K, and there were only thirty-eight neutrons for every sixty-two protons. About a third of a second after time zero, neutrinos ceased to interact with ordinary matter, except through gravity and the ordinary kinetic effect of collisions.

By the time the Universe had cooled to $10^{10}$ K (10 billion K), 1.1 seconds after time zero, its density was down to just 380,000 times the density of water, and the balance between protons and neutrons had shifted further, with twenty-four neutrons for every seventy-six protons. By the time the Universe had cooled to 3 billion K, 13.8 seconds after time zero, nuclei of deuterium, each containing one proton and one neutron, began to form, but were soon knocked apart by collisions with other particles. Only 17 per cent of the nuclear particles (nucleons) were now left in the form of neutrons.

Three minutes and two seconds after time zero, the Universe had cooled to 1 billion K, only seventy times hotter than the centre of the Sun is today. The proportion of neutrons was down to 14 per cent, but they were saved from disappearing entirely from the scene because the temperature had at last fallen to the point where nuclei of deuterium and helium could be formed and stick together in spite of collisions with other particles.

It was at this epoch, during the fourth minute after time zero, that the reactions outlined by Gamow and his colleagues in the 1940s, and refined by Fred Hoyle and others in the 1960s, took place, locking up the remaining neutrons in helium nuclei. The proportion of the total mass of nucleons converted into helium is twice the abundance of neutrons at the time, because each nucleus of helium (helium-4) contains two protons as well as two neutrons. Four minutes after time zero, the process was complete, with just under 25 per cent of the nuclear material converted into helium, and the rest left behind as lone protons – hydrogen nuclei.

A little more than half an hour after time zero, all of the positrons in the Universe had annihilated with almost all of the electrons – again with one in a billion left over, matching the number of protons – to produce the background radiation proper. The temperature was down to 300 million K, and the density was only 10 per cent of that of water. But the Universe was still too hot for stable atoms to form; as soon as a nucleus latches on to an electron under those conditions, the electron is knocked away by an energetic photon of the background radiation.

This interaction between electrons and photons continued for 300,000 years, until the Universe had cooled to 6,000 K (roughly the temperature at the surface of the Sun) and the photons were becoming too weak to knock electrons off atoms. At this point (actually, over the next few hundred thousand years) the background radiation decoupled, and had no more significant interaction with matter. But by this time the gravitational influence of all the matter in the Universe had built on the tiny irregularities present from the early stages of the Big Bang, tugging the hydrogen and helium into great clouds that left their imprint on the background radiation, and which would one day collapse to form clusters of galaxies. The Big Bang was over, and the Universe was left to expand relatively quietly, cooling as it does so, and expanding ever more slowly as gravity tries to pull it back together.

# Further Reading

George Gamow, *The Creation of the Universe* (Viking, 1952)
John Gribbin, *In the Beginning* (Viking, 1993)
John Gribbin, *Companion to the Cosmos* (Weidenfeld & Nicolson, 1996)
Laurie John (ed.), *Cosmology Now* (BBC Publications, 1973)
Martin Rees, *Before the Beginning* (Simon & Schuster, 1997)
Lee Smolin, *The Life of the Cosmos* (Weidenfeld & Nicolson, 1997)
Steven Weinberg, *The First Three Minutes* (Deutsch, 1977)

# PREDICTIONS

Asia and the Pacific
Climate
Cosmology
Crime and Punishment
Disease
Europe
Family
Genetic Manipulation
Great Powers
India
Liberty
Media
Men
Middle East
Mind
Moral Values
Population
Religion
Superhighways
Terrorism
Theatre
USA
Warfare
Women

Rabbits they say
Are very scarce to-day
My diagnosis?
Myxamatousis.

NON RABBIT
ELEPHANT.

Terence Newt
Wore a Giant boot
Jammed down over his head
And he kept it there
{with his ears and hair}
Until the day he was dead.
But when his wife removed the boot,
She discovered to her horror!
It was not the head of Terence Newt
But three other men. Tom Daft an apprentice butcher,. Cyril hunge a Mechanic and Arthur Woggs. Dent

**Terence Newt**
**Wore a giant boot**
**Jammed down over his head**
**And he kept it there**
**With his ears and hair**
**Until the day he was dead.**
**But when his wife removed the boot,**
**She discovered to her horror!**
**It was not the head of Terence Newt**
**But three other men. Tom Daft an apprentice**
**butcher, Cyril Lunge a Mechanic, and**
**Arthur Woggs, Dentist.**

# Hopeless Love

It is a hopeless time.
You are spring
And I –
Am Autumn.
Love can't close the gap.
And if it could – if it did –
One day you will be Autumn
And I Winter – and after that?

Midnight
Bournemouth
April 1967

A Boarding house
in Christchurch Rd

The Jet Plane –!
Brilliant – time saving
The new Icarus – with Tungsten Wings
And for me –
It just takes me from one unhappiness to another –

## The Young Soldiers

Why are they lying in some distant land
Why did they go, did they understand?
Young men they were
Young men they stay
But why did we send them away, away?

written during Korean War
March 30th 1955

Millygannmack

Got a Picture of a Gnu -
Cut it out. place it over a
news paper. Caption. The
10' o'clock Gnus.

## Christmas Morning

A little girl called Silé Javotte
Said 'Look at the lovely presents I've got'.
A little girl in Biafra said
'Oh! What a lovely slice of bread.

## Samson and Delilah

As he pushed the pillars apart
Samson was appalled
For just before the palace fell in
Delilah said 'He's bald!'.

# Time Gentlemen

A-tick- A-tock.
Goes grandfather clock,
All through the night.
And, every hour
With tremendous power
The clock would start to chime.

And may,
I say,
Its a noisy way,
For it,
to tell,
the Time.

## A Nose

## A World War II Nose

My nose, my nose lived dangerously
Its courage was no stunt!
And during the war in Germany
It was always out in front!

Yet when the battle was o'er
And we'd defeated the Hun
Suddenly, for no reason at all
My nose started to run.

## These things called

The human face is something that
Hangs downwards from a thing called hat
And when the hat is raised, it's said
It shows a hairy thing called head.
Now I would rather cover face
And strike it full on with a mace.

Mermaid Theatre. 20 Dec. 1967

## Brave New World

Twinkle Twinkle, little star
How I wonder what you are
Up above the sky so high
Like a diamond in the sky

Twinkle Twinkle, little star
I've just found out what you ar
A lump of rusting rocket case
A rubbish tip - in outer space.

Dear Reader!

Dublin
Nov. 1967

Human beings will become so used to being crushed together that when they are on their own, they will suffer withdrawal symptoms. "Doctor—I've got to get into a crowded train soon or I'll go mad". So, special N.H.S. assimilated rush hour trains will be run every other Sunday for patients. At 9 o'clock on that morning, thousands of victims will crowd platforms throughout England, where great electrically powered Crowd Compressors will crush hundreds of writhing humans into trains, until their eyes stand out under the strain, then, even more wretches are forced in by smearing them with vaseline and sliding them in sideways between legs of standing passengers. The doors close—any bits of clothing, ears or fingers are snipped off. To add to the sufferers' relief great clouds of stale cigarette smoke are pumped into the carriages. The patients start to cough, laugh and talk. They're feeling better already. But more happiness is on the way. The train reaches 80 m.p.h., at the next station the driver slams the brakes on shooting all the victims up to one end of the carriages. Immediately the doors open, and great compressed air tubes loaded with up to 100 passengers are fired into the empty spaces, this goes on until the rubber roofs of the carriages give upwards, and the lumps you see are yet a second layer of grateful patients. Off goes the train, and one sees the relief on the travellers' faces. Who wants LSD when you can get this? Ah! you say, the train can't possibly take any more. Wrong! At the next stop the train is sprayed with a powerful adhesive glue, and fresh passengers stuck to the outside, and so, crushed to pulp, pop-eyed and coughing blood, the train carries out its work of mercy. Those who are worried about their children's future in the 20th century need not fear. We are prepared.

TAKE ME TO YOUR LEDA
Spike Milligan

Mr. Timothy Pringle
Lived on his own
As he was single.
Returning from work
In the evening gloom
He found an elephant
In his room.
It had a label
Round its neck
"My name is Doris
Eileen Beck".
Even if the name was Jim
It didn't really help poor Tim.
Is that elephant a her or he?
Asked Mr. Screws (the landlady)
Tim said "It's a female elephant, why?"
"No women in rooms" was the stern reply

Mr Timothy Pringle
Lived on his own
As he was single.
Returing from work
In the evening gloom
He found an Elephant
In his room
It had a label
Round its neck
"My name is Doris
Eileen Beck".
Even if the name was Jim
It didn't really help poor Tim.
Is that Elephant a her or he?
Asked Mr Screws [his landlady]
Tim said Its a female elephant why?
"No women in rooms' was the stern reply

## VALUES '67

Pass by Citizen:
   don't look left or right,
keep those drip dry eyes straight ahead.
A tree? Chop it down.
      they're a danger to lightning
Pansies, calling for water?
      Let 'em die — the queer bastards!
Seek comfort in the scarlet plastic
            labour saving rose
Fresh with the fragrance of Daz.
Sunday. Pray citizen:
   pray that no rain will fall
      on your newly polished
               four wheeled
                  God.

Envoi. Beauty is in the eye of the beholder.
Get it out — with Optrex.

On a train to Liverpool, Easter Monday 1967.

Phillip le Barr
Was knocked down by a car
On the road to Mandalay
He was knocked down again
By a dust cart in Spain
And again, in Zanzibar
So,
He travelled at night
In the pale moon-light
Away from the traffic's growl
But terrible luck
He was hit by a duck
Driven by—an owl.

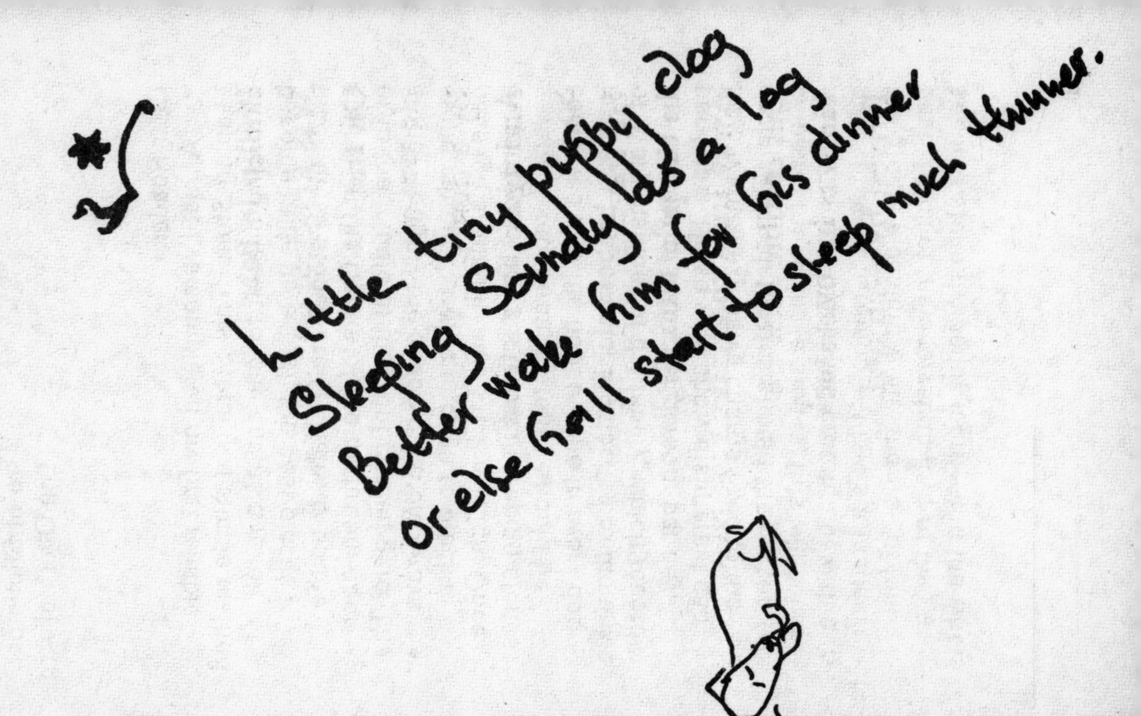

Little tiny puppy dog
Sleeping soundly as a log
Better wake him for his dinner
Or else he'll start to sleep much thinner.

Letters to Anyone

3rd of Sliptimber
The Olleth of Arg

Dear Reader,

At last I've completed my first reliable Leaping Stone. This Stone is a simple method of forcing oneself to leap and as a result become quite an expert, the best places to place these stones are (1) Halfway up the stairs—in the Centre of the Bathroom door—or in the middle of the front garden gate. The stone—roughly the shape of an old Grave stone, should be wide enough to block a passage around it, and 3 ft. 2 in. high. Once a client has shown interest in leaping stones several demonstration stones should be cemented into his house—one in every door in fact, a Leapo-meter is attached to the ankle of every member of the family, which records the number of leaps per person, per day, per-haps. Those who show disinterest can have a small explosive charge fixed to the groin, which detonates should the person try climbing round these stones, this will cause many a smoke blackened crutch, but with our new spray-on 'Crutcho'—a few squirts leave the groins gleaming white and free of foul-pest. Think of the enervating joys of the Leaping stones! Sunday morn—and the whole household rings with shouts of Hoi Hup! Ho la! Grannies—uncles—Mothers—cripples —all leaping merrily from one room to another —ah, there's true happiness. We have high hopes that more progressive young politicians with an eye to eliminating senile M.P.s, intend to have a 'Great Westminster Leaping Stone' that will be placed dead centre of the Great entrance doors on the opening day of Parliament—those failing will of course be debarred—though they can claim 'A brethren assisted leap'—this means that two decrepit M.P.s of the same party, can try and assist the failed member over the leaping stone by applying hot pokers to the seat, thus the smell of scorched seats, burning hairs and screams, can bring a touch of colour to an otherwise dull occasion. I don't know when I will post this letter, I might deliver it tomorrow by hand, ankle, foot and clenched elbow.

As ever,
Spike.

It's in
here
somewhere

2

Oh the Wiggley-Woggley men
They don't get up till ten
They run about
Then give shout
And go back to bed again!

Oh the Wiggley-Woggley men
They don't get up till ten
They run about
Then give a shout
And back to bed again!

An Ear passed me
the other day
And silently
went on its way

I wonder who
that ear can be
And has it ever
heard of me.

Take us to you leider.

## Manic Depression

St. Lukes Wing
Woodside Hospital
Psychiatric Wing 1953 -

The pain is too much
A thousand grim winters
grow in my head.
In my ears
the sound of the
coming dead
All seasons, all same
all living
all pain
No opiate to lock still
my senses.
Only left, the body locked tenser.

December 1960.

Dec. 1960

Manic Depression

The pain is too much
A thousand grim winters
grow in my head.

In my ears
the sound of the
coming dead

All seasons, all same
all living
all pain
No opiate to lock still
my senses.

Only left, the body locked tense.

Norrington Blitt
Ate aught but grit
Ate aught but grit and mussels
But when he got there
The cupboard was bare
Save a sack of sprouts—
From Brussels, or was it Oldham?
No—a tree fell on him.
Or was—

# Freedom

A bird in flight,
her wings spread wide
Is the soul of man
with bonds untied
Beyond the plough
the spade, the God
A bird flies in
the face of God,
Yet I with reason
bright as day
Forever tread
the earthbound clay.

## PARIS PAREE

Written in Paris when
I went last time.

Paris! Paree! What pictures of gaiety those two cities conjour up, down, and sideways, Paris, city of Napoleon, the Revolution, the Mob, the blood, the head rolling. Alas, those happy days are gone, yet, Paris, the Queen of cities calls us all. Last week it called me, 'Cooee!' it said and I responded. Travel allowance being only £50, I saved by taking sandwiches and a Thermos of Tomato Sauce. I saved further on the air fare by travelling second class non-return tourist night flight, all you had to do was sign a Secret Enoch Powell form saying you were an undesirable coloured alien with uncurable bed-wetting. At the airport there was the carefully disguised panic rush to get the back seats in the plane. On take-off I fastened my safety belt, read 'How to inflate Life Belt', swallowed a boiled sweet, made the sign of the cross and read the Times. One hour later coming in to Orly I fastened my copy of the Times, made the sign of the seat belt, swallowed my boiled life belt and inflated myself for landing. Through to Customs and out! At the airport my taxi drew up in a cloud of Garlic, and the driver leapt out and gesticulated in a corner.

Arriving at the Hotel, the porter raised his hat and lowered his trousers. Real French hospitality! The Hotel had been built in 1803—in 1804 they added a west wing and in 1819 it flew away. Next morning I was up at the crack of noon shouting "Apres moi le deluge" and whistling Toulouse Lautrec, I hurriedly swallowed a breakfast of porridge and frogs and a steaming bidet of coffee. I next joined a crowd of impoverished British tourists on the 30 centimes all-in English punishment Tour. A great herd of us assembled at the Place du Concord, from there we were force-marched to the Notre Dame, beaten with sticks and made to climb the great Bell Tower. Sheer physical agony! On the way up we passed many who had perished in the attempt and never made it. Fancy! 600 steps! No wonder Quasi Modo had a hump on his back when he got to the top! From the top I took several lovely photos of the Eiffel Tower. At Midday, we were led to a Cafe 'Le Gogo Plastique', the establishment bore the indelible stamp of the British tourists—

**Menu**

'Escargots and mash . . .'
'Bisque d'Homard, bread and butter'
'Pate de Fois Gras and Chips'
'Lobster Thermidor, 2 Veg., Boiled Pots. etc.'
'Crepe Suzette Flambe and Custard'

The lady next to me had Frogs' legs, her friend's weren't much better. It's all that walking, I suppose. I was served by a waiter who made it perfectly clear he held me personally responsible for a) The loss of Algeria b) Waterloo c) Edith Piaff. Just so they didn't think I was an oaf, I ordered the whole meal in French—I was brought a hip-bath, a silk tie, a coloured pencil and a small clockwork Virgin Mary that whistled Ave Maria every hour, made in Hong Kong. I spent the rest of the afternoon sketching the beautiful Eiffel Tower. There's always something to do in Paris! Carefully following my Baedecker's Paris I walked up the hill to the Cemetery of Pierre Lachaze, I saw the very spot where Moulin Rouge lay buried, and above me gleaming white was the Sacred Cur, now used as a church. I had been walking some three hours and as a quick calculation showed me that I was exactly six miles from the lovely Eiffel Tower, I took a taxi back to the Pension; to my horror he asked for 13 francs, I was about to have a show down with him, but, rather than ruin the evening, I paid him. It ruined the evening. I freshened up in my room, taking a shower and a foot-bath in a very low basin with a rather dangerous water jet that took me completely by surprise. The evening would be dedicated to Art, I always wanted to see the French Impressionists so I booked for the Folie Bergere where a man was doing imitations of Maurice Chevalier, Josephine Baker, and many others. What a show! Women uncovered from the waist up, and yet there was a cover charge! Watching women with naked bosoms is unsettling, but eventually they grow on you. If they grew on me I'd go to the pictures alone. The Grand Finale was called 'Salute les Anglais', the band played a Pop version of God Save the Queen while a French queer wearing a Prince Philip mask juggled with three Plastic Busts of the heir presumptive. It was good to see that we were still a country to be respected. If only we had their Eiffel Tower there'd be no stopping us.

29

Painted in 1866 (canvas 11" ×

Blot on the Landscape

George Melley
Had such a fat Belly
He couldn't get near the Telly
So he had to go
And listen to the Radio.

All the ravel-avel tumble-umblings — all the high sorrows —
All going all giving all taking all doing. Which, what
When where how — worse still. Who? Who's next? —
I know — it's me — it's always me; time for it now sir —
Now son — now — now — NOW it's time for it. Yes.
Now the drop down starts — sitting, standing, kneeling
Lying — the drop starts — down down — it's always down —
Some time down stops — but it only stops at down — then
Off it goes again — down, down, down, into the unanswerable.

Over Elba — September, 196
Drunk

Skeleton of Prehistoric Car

Brontacycle

Said the mother Tern
to her baby Tern
Would you like a brother?
Said baby Tern
to mother Tern
Yes
One good Tern deserves another.

Sent to Chiselhurst & Sidcup Colleg

I see those two have fallen out with each other.

9 Orme Court,
Bayswater,
W.2.
13 June, 1963.

Sir,

Whereas the Bishop of Southwark is to be complimented on his speech (Times, 13 June '63) when he speaks of "corruption in high places", what he should have said was "corruption in all places". To point the finger of indictment at one or two persons, is almost laughable when one considers the extent of moral corruption in this country as a whole, and this corruption feeds on the licence it is allowed by the feeble Church and Parliamentary laws, which almost condone it. Pornography, which is the greatest inciter of immorality, is not only rife, but actually bestowed, in some spheres, as 'art', by the intellectuals, which is quickly exploited by all commercial enterprises. Films of rape, murder, sex, debasement, are now 'la mode', to criticise them is to be 'old hat'. The X certificate is box-office bonanza. Even innocent comedies have publicity that is aimed at sex. As a parent, I find it increasingly difficult to take my children to see a suitable film. Bookstalls in all our major cities groan under the weight of pornographic 'literature', Men's 'clubs' doing nothing but strip-shows, drinking cluhs show films that are laughingly called

'Naked but free' etc., etc. Photographic pornography, using safe Box numbers, is a million pound trade. In the light of this I fear Dr. Stockwood's address will have little or no effect at all, yet, it is in religious laxity that the seeds of immorality grow, and, at this critical stage in history, never has the Christian church been so inactive and indifferent to the massive danger to Christianity. Take China as an example, the Christians were in China 500 years before Communism. Today in China, Christianity hardly exists, why? Parrot-indoctrination; give the native a vest, a crucifix, one chorus of 'Onward Christian Soldiers', and he's ticked off as being a Christian, and still Christians are churned out as tho' from a mould, size is no substitute for quality. I myself was baptised a Catholic, and I still don't know what it is all about, nobody ever bothered to teach me. Year after year I listen for a message of enlightenment from the pulpit, but no, the Gospels, the Epistles, are repeated ad nauseum, but of contemporary guidance there is nothing. Going round saying "And the Lord said unto Moses" won't get us anywhere, Jesus didn't talk about throwing the money lenders out of the Temple, he did it, then talked. If we are to stop the moral rot, we must act, we must indict, we must mention the offenders by name and not hide for fear of libel, the truth is all that matters, right now this country is not geared to accept it.

Consider, that England, France, Italy, West Germany and America are where pornography abounds, and these are the countries dedicated to preserve Christianity from the Godless Russia who has no pornography.

Spike Milligan

## A little poem for Sean

There was a young boy called Sean
Who sat on the edge of a lawn
His knees went crack!
He fell on his back
And regretted the day he was born.

I sent my legs out for a walk
To keep them strong and fit
They would not go without me
So I've made the b . . . . . s sit.

THINK OF THE MONEY YOU'LL SAVE IN TRANSPORT.

YOU'LL FIND YOU'LL BE ABLE TO GET AROUN MUCH QUICKER.

Cautionary Letter

Green Bonk

Dear reader, the worst has happened. Brown Bonk has struck in the Quantocks. Worse still, attacks of Green Bonk have been reported coming from the Urals; an Armenian fruit shepherd was driving a herd of Apple trees to water when the Brown Bonk laid him low. The first case of our own Brown Bonk was at Catford Labour Exchange. Mr. Ted Naffs was being given his certificate for 21 years of devoted unemployment, when, yes! Brown Bonk! Smoke started to issue from Mr. Naffs' mouth, a scream of agony showed his teeth to be molten white. He was rushed to a blacksmith who removed his ring of confidence. Toothless with Brown Bonk, Mr. Naffs was given a pair of N.H.S. electric teeth, for high speed eating, a boon to the aged and infirm. Leave the food and teeth by your bed at night and they do your eating while you sleep. Alas, Mr. Naffs' house, was all Gas. An application had to be made for a set of North Sea High Speed Gas-operated Teeth. While he was waiting for delivery Mr. Naffs stupidly plugged into the great overhead cables of the National Electric grid which ran over his cottage. As he switched on

100,000,000 volts shot into his 240 volt teeth, a brilliant flash of magnesium, and his dentures started to chew at round about the speed of light, Mr. Naffs' head became a white blurr as his teeth ate the porridge—the plate—knife—fork—spoon—table mat—the table—the chairs—the dog—the cat—two budgies—a Welsh dresser, he was half way through a brick wall when the annual power failure saved him. Beware all of you, Brown Bonk is with us, at the first sign of smouldering teeth. . . . . . . . . . . . Write to the fire brigade.

I had a Dongee
Who would not speak
He wouldnt hop
He would'nt creep
He would'nt walk
He would'nt leap
He would'nt wake
He would'nt sleep
He would'nt shout
He would'nt squeak
He would'nt look
He would'nt peep
He would'nt wag
his Dongee tail.
I think this Dongees
going to fail.

**If Robert Graves**
**misbehaves**
**It's the Torjca**
**Majorca**

Ooer the hairy Sporran!
Harris Tweedie in the Glen!
Aw braw a cumming oot!!
Wi Sixty thousand men!!!
nd Och! yer maither tellin!!!...
Awa wi ye wee Ben!
An taka swirrl and Dream Goin'
n the News at 5 to 10!!!!

**The sayings of Mrs. Doris Reach**
**of 23 The Irons, Cleethorpes, Herts.**

1. Ups a Daisy
2. Save the string.
3. There's some in the tin on the Mantlepiece.
4. They should never let them in the country.
5. Ups a Daisy
6. I had an Aunt who was like that.
7. Mine's Brown.
8. Just a small one then.
9. Ups a Daisy.
10. It comes off with Turps.
11. He knows every word you say.
12. Ups a Daisy.
13. What about a nice....................
14. Ups a Daisy.

Editor. Rag Mag., Gloucester College of Education.

You say your mag. is in aid of mental health! Dear Lad, there's no such thing, if there was anybody in position of power with any semblance of mental health do you think the world would be in this bloody mess? Young minds at risk is different. Anybody with a young mind is taking a risk—young means fresh—unsullied, ready to be gobbled up in an adult world bringing the young into visionless world of adults, like all our leaders. Their world is dead—dead—dead, and my God, that's why it stinks! They look at youth in horror—and say 'They are having a revolution', but what do they want? I say they don't know what they want, but they know what they don't want, and that is, the repetition of the past mistakes, towards which the adult old order is still heading. War—armistice—building up to pre-war standards—capitalism—labour—crisis—war and so on. I digress.
Mental Health. I have had five nervous breakdowns—and all the medics gave me was medicine—tablets—but no love or any attempt at involvement, in this respect I might as well have been a fish in a bowl. The mentally ill

need LOVE, UNDERSTANDING—TOLERANCE, as yet unobtainable on the N.H.S. or the private world of psychiatry, but tablets, yes, and a bill for £5 5. 0. a visit—if they know who you are it's £10 10. 0. a visit—the increased fee has an immediate depressing effect—so you come out worse than you went in.
As yet, I have not been cured, patched up via chemicals, yes. Letter unfinished, but I've run out of time—sorry!

Regards,
Spike

Once there was a girl
Who grew roses in my head
Made a paradise of bed
Yet not a word was said.

# My Street

Pass by
Pull the blinds
Nothing happened here to-day

Douse the light
lock the door.
Nothings happened.

Walk the street
turn the corner
Nothing there. forever more

J

What was it –
so quick.
So Unthought
Yet – Alright
It was'nt me
It was'nt you
Yet it was —
Before "
After "
During
We were waiting
for the
answer to
one plus one

What was it -
So quick
So unthought
Yet - all right?
It was it me
or you -
them - ~~it~~ was
Before -
After
During - surely the same
We are just all the answer
[illegible]

Ronnie Scotts Club

What was it?
So quick
So unthought
Yet—Alright
It wasn't me
It wasn't you
Yet it was—
Before
  After
    During.

We were looking for the
answer to one plus one.

Ronnie Scotts Club.

*"No, no, try ASPIRIN."*

Dear Lads,

I am so overwhelmed with work at the moment, and by this, I don't mean the sheer monetary kind: I also try and have an interest in humanity.

Anyhow, I just haven't got time for an article. However! I would like to state that I have been aware of the coming dissent among students for the last ten years against a world, which has become archaic in ideas, and the extension of those ideas.

The normal channels left open to citizens to complain are crammed to the brink, and normally no individual citizen can get anything changed under the present system, unless he is willing to spend £10,000 and go to litigation, that, or set fire to himself outside No. 10 Downing Street, which would, in turn, only result in a bucket of water from the Prime Minister.

Whereas, students are not at all clear as to what they want, they certainly know what they don't want.

The old world and its standard formula of democracy are spent forces, who can only go on repeating past mistakes on the same democratic basis, among these we have over-population, starvation, war, political unrest, and archaic laws which themselves are evil. One only has to investigate the modern divorce law to discover to what disgusting levels human beings are reduced, to obtain a divorce.

Religion has become unchristian, the Christ that lived would certainly look extremely embarrassed and out of place in the Vatican. Christian morals have changed to such an extent that people are now on the verge of accepting debasement as some form of art. Jesus chose a donkey. Top clergy use Austin Princesses!

I myself am not a prude, but I am finding it difficult to find a film, a book, or a magazine that does not use the sex act as the basis of world art form.

All this is being seen very clearly by student bodies around the world, and through the extraordinary combustion exploding at the same time.

Unfortunately, the old guard are so firmly entrenched and the system so protected by armed forces, police and lackies, that only civil war can bring about a change.

Modern student thinking is violently opposed to the use of arms (or at least it should be), but student bodies must keep up the pressure for the rest of their lives to try and influence their wives, and children and children's children into thinking of a new method for man to live on this earth. Governments can start by solving the most important problem in the world, over-population, which in itself leads to the lowering of all man's standards.

I haven't time to say much more, but let us say that the old world has got to change, and by the old world I also mean these idiots who want to get to the moon, when man hasn't really got to the earth.

Sincerely,
Spike Milligan

P.S. I keep fighting.

## Its little God,

When my daddys in the bath
I knock upon the door
'Whos dat' he says
But I dont say and then?
.I knock some more,
Whos dad' hes says' Whos dal aga
And he must think its odd
For when at last I answer him
'Its me' I say "Its God!"

**He swung to and fro**
**Then  3 and fro**
**4 and fro**
**5 and fro**
**And finally 6 and fro.**

## SUN HELMET

A pleasant three degrees below zero wind was blowing. The early morning Londoners shivered through the bitingly cold rush hour. Among them was a bowler-hatted Mr. Oliver Thrigg. The first snow of summer was starting to fall as he joined his 'AA members only' bus cue. Glancing to a bus que opposite (it was a different que to his cue, as the spelling proves), and what he saw shook him to his foundation garment. There, in the que opposite, was a man wearing a sun helmet, eccentricity yes, but this fellow didn't have a stamp of a genuine eccentric, no, fellow looked far too normal! Curiosity got the upper hand, crossing the road he killed a cat. Once across he joined the que and left his on the other side. The man in the sun-helmeted man caught a 31A bus, Mr. Thrigg signalled a passing 49A. "Follow that bus" he told the driver. "Anywhere but Cuba" said the driver. At Victoria Station the sun-helmeted man booked to Southampton, as did Mr. Thrigg, who kept him under surveyaliance until they reached Southampton, where by now the snow was 3 foot deep, which explained the absence of dwarfs in the street. The man continued to wear his sun helmet. "Why, Why, Why" said Thrigg whose curiosity had killed another nine cats, making a grand total of one. "I must follow this man etc." The man booked aboard the Onion Castle and was handed £10 and an oar (Assisted Passage they call it). The ship headed south, and, so did Mr. Thrigg and his enigma, which he used for colonic irrigation. During the whole trip the man appeared at all times in a sun helmet. Several or eightal times he was almost tempted to ask the man his secret. But no, as Thrigg was travelling steerage and the man 1st Class, plus the fact it was a special Non-fraternising Apartheid Cruise, no contact was possible. On the 12th of Iptomber the ship docked at Cape Town. Even though Thrigg got through Customs and Bribes at speed, he just missed the Sun Helmet as he drove off in a taxi. Thrigg flagged down an old cripple Negro driver "Follow that Sun Helmet" he said jumping on the nigger's back. (The change from Negro to Nigger denotes change from UK to SA soil.) Several times Thrigg let the nigger stand in his bucket of portable UK soil so he could be called Negro. To cut the story short, Mr. Thrigg used scissors and cornered the man in the middle of the Sahara. The heat was intolerable as Thrigg walked up and said "Why are you wearing that sun helmet?" "Because said the man, pointing at a 113° thermometer in the shade "The sun man! This protects the head." "I see" said Mr. Thrigg. "Well I better be off, I'm late for work." As he departed for the caravan que, the man in the sun helmet spotted him. "Good God, a man wearing a bowler hat! A bowler hat? Here, in the Sahara? I must find out why," he thought as he joined the caravan cue behind Mr. Thrigg.

Say When ——

## Titikaka

The magic green lake
that fell from the sky,
quenched a burning mountain's throat
and sent a fire king
into untimeable slumber.

On a plane over Mexico. Sept. 19

This should settle it once and for all.

## Morning

Chamfers of White
Light
Shaft from the cheap 20th Century glass
Flared on my bed
Red
Blanket barks its reflection to the ear in
my e

And so it
bit
By bit pulls together the strings of mornin

January 4 at 'Old P

hashing – hew

At the third stroke it will be 3.29

The 12th June
is very soon
for there is a party on that Day
and most liky I be going away.

Silve.
June 6

The 18th June
Is very soon
For there is a party
  on that day
And most likely I be
  going away.

Silé. June 1967

**A bottle of Graves**
**or Graves**
**I ordered in half**
**or halves**

**How teeny teeny wee**
**Is the teeny little flea.**
**But last night in my hotel**
**He made me scratch like merry hell!**

Foreign Office,
June 23, 1897

Sir:

I have the honour to acknowledge the receipt of your Note of the 19th instant, in which you transmit a copy of your Credentials as Special Envoy from the Argentine Republic on the occasion of the celebration of the Sixtieth Anniversary of the Queen's Accession to the Throne.

I have the honour to be,
with the highest consideration,
Sir,
Your most obedient
humble servant,

(FOR THE MARQUIS OF SALISBURY)
F. H. Villier

Monsieur Florencio Dominguez
&c., &c., &c.

## THE TALE OF FRANCIS PAW

by Margaret and Jack Hobbs

There was a cat called Francis Paw
Who lived behind the kitchen door
He was old and brown
And tumbledown
But his fur was long and silky
He liked an egg for breakfast
And some tea if it was milky.

He often sat in an old silk hat
And talked to me of this and that
He liked a joke
And his voice would croak
As he told me of his youth.

He used to croon
To the harvest moon
And sing a roundelay
Of his Uncle Fred
Who is long since dead
Who went to Botany Bay.

He went aboard a whaling boat
Which foundered in a gale
With feline craft
He clung to a raft
With the end of his scraggy tail.

At last he reached an island fair
And struggled to the shore
Where under a tree
Stood Sam Macgee
Staring at Frederick Paw.

'Yo Ho' he cried
'The cook has died
'What can we have to eat?'
'You look a tasty morsel, lad,
'Except for your smelly feet.'

Fred's ears went flat
'I'll catch a rat'
He said in trembling tones
There isn't very much of me
I'm only skin and bones.

'I would' said Sam
'Prefer a ham'
But I'll give you an hour
To catch a rat that's big and fat
And cook it rolled in flour.

Fred caught a rat
And that was that
And rescue soon was nigh
In the shape of a Chinese sailor
By the name of Wun Flung Hi.

Wun Flung Hi
Had a drooping eye
And a most unsavoury crew
They hit poor Fred
And Sam on the head
And put 'em in Hold No. 2.

They sat in the hold
Through the long dark night
Wondering what to do
When the hatch was raised
And down there gazed
The face of Fu Manchu.

Fu Manchu was a prisoner too
For a ransom so they say
But he'd opened the lock
With a piece of a clock
And planned a getaway.

Fred and Sam and Fu Manchu
Took a boat and sailed away.
And when the dawn
Of the day was born
They landed at Botany Bay.

What happened then
I chose to ask
Of my friend, young Francis Paw
That's another tale
He yawned and fell
Asleep by the kitchen door.